DISASTER ON OUR DOORSTEP

How Juneau responded to the worst shipwreck in North Pacific history

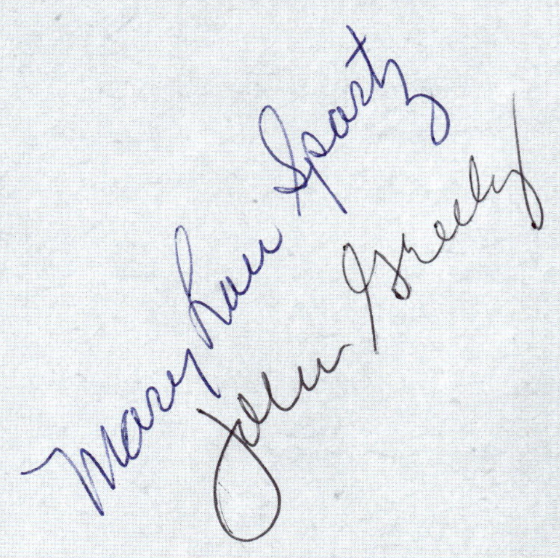

BY
MaryLou Spartz
WITH
John Greely

DISASTER ON OUR DOORSTEP

ISBN: 978-1-57833-710-1

First Printing August 2018

Book Design: Crystal Burrell, Todd Communications

Published by:
MaryLou Spartz
P.O. Box 20211
Juneau, AK 99802
John_greely@hotmail.com

Distributed by:

Todd Communications
5452 Jenkins Drive Suite 3
Juneau, Alaska 99801
Phone: (907) 274-TODD (8633) • Fax: (907) 929-5550
with other offices in Anchorage and Fairbanks, Alaska
WWW.ALASKABOOKSANDCALENDARS.COM • sales@toddcom.com

Front and back cover paintings by Dan Fruits

TABLE OF CONTENTS

THANK YOU

The authors would like to express their appreciation for assistance from the staff of the Alaska State Historical Library and the Juneau-Douglas City Museum, where the past is kept alive for all of us. The staffs of the Maritime Museum of British Columbia and the Royal British Columbia Museum Archives were also very responsive to our information requests.

We thank artist Dan Fruits of Juneau for his generosity in allowing use of his watercolors for this book. And we sincerely appreciate the suggestions of Sarah Isto that improved the text.

The community of Juneau hosted events throughout 2018 around the 100th anniversary of the sinking. Members of the Princess Sophia Centennial Commemorative Year Planning Committee include Annette Smith, Barbara Burnett, Peggy Cartmill, Carl Brodersen , Don Burford, Addison Field, Fred Thorsteinson, Claire Imamura, Kathy Ruddy, Katy Giorgio, Peter Naoroz, Robert Barr, Terry Brenner, Tom Dawson and Todd Hunt. Thanks to Fireweed Place staff for providing a meeting place for the committee.

The authors would be remiss if we didn't also mention the original opera created for the anniversary. "The Princess Sophia" was conceived and written by Jim Simard, David Hunsaker and Emerson Eades and scheduled for debut in Juneau on October 25, 2018.

Part I
Introduction

Image courtesy of the Juneau-Douglas City Museum, JDCM 2015.29

The SS Princess Sophia *is pictured at the Juneau waterfront in 1918. The Canadian Pacific ship was part of a fleet serving the Pacific Coast between Vancouver and Skagway.*

When I was growing up on Twelfth Street, our back yard bordered Evergreen Cemetery, Juneau's burial ground. I was a curious eleven-year-old kid and considered it, as did most neighborhood kids, a place to play. I delighted in wandering around among the tombstones, reading the names and dates and imagining what the people buried there might have been like.

Photo courtesy of the University of Alaska Collections

Walter & Frances Wells Harper

One grave really made me wonder. Two people, Walter Harper, and his wife, Frances Wells, were buried side-by-side under a large marble slab that told of how they perished when the *SS Princess Sophia* went down on Vanderbilt Reef. No one had ever talked about a big ship sinking in Lynn Canal with 353 people on board. So I asked my Dad about it.

Dad told me the story of a Canadian ship traveling south from Skagway to Vancouver, British Columbia in 1918 when she was pounded by a huge October storm. She ran aground at night on the reef about halfway between Skagway and Juneau. Smaller boats from Juneau hurried to offer aid. They were willing to take the passengers to safety, but the captain of the *Princess Sophia* refused all offers of help.

The monster storm continued and the ship remained fast on the reef for about a day and a half. On the second afternoon, the boats had to run for protection, promising to return at first light. When they did return, the rescuers found only the mast sticking out of the water and all lives lost. Dad finished by saying shortly after the disaster a dog was found in Auke Bay covered with oil and thought to have come from the wreck. It was adopted by a family and lived out its days with them.

I was glad that some living thing had survived.

It seemed like too large a story for the small towns of Southeast Alaska, where people knew everyone and death came in small doses, one accident or a cave-in at a time.

In 1918, Juneau and Douglas were approaching their 30th birthdays as gold towns. A phone system and a ferry across Gastineau Channel connected the two towns. But what held the citizens together was the shared experience of everyday life as seen in newspaper articles about ladies' tea parties, and about people who had just arrived or departed on steamships such as the *Princess Sophia*, and increasingly that year, news of World War I.

America had entered the war in the spring of 1917, and some two million young men enlisted nationwide. About 50,000 of them (81 from Alaska) died before the war ended in November 1918. Many of these Alaska men succumbed to influenza contracted at military training sites before they even left the country. In the fall of 1918, schools were closed in Juneau to prevent the flu from spreading.

The largest employer in Alaska, the Treadwell Alaska Gold Mine, had collapsed into the channel in front of Juneau in 1917, without fatalities but taking with it about 1,000 mine and mill jobs. However, residents remained optimistic that a new mine would rise at the competing Alaska-Juneau operation.

But nothing in Juneau's experience would rival the sinking of the *Princess Sophia*, or as one newspaper editorial called it, "Our Appalling Catastrophe". This book chronicles how Juneau and neighboring towns responded to the disaster on their doorstep, from rescue and recovery efforts to the return of remains to families for burial at a final resting place at Evergreen Cemetery.

Flooding caused the collapse of the Treadwell Alaska Gold Mine in 1917. No one died in the flooding, but Alaska's largest private employer at the time never recovered, forcing about 1,000 miners and mill workers to find other jobs.

PART II

THE ACCIDENT AND RESCUE EFFORT

Photo courtesy of Joel Bennett

Vanderbilt Reef is shown at low tide. The reef is the tip of a submarine mountain estimated at 1,000 feet in height. It is frequently submerged at high tide.

Photo ASL-P523-09 courtesy of the Alaska State Library Historical Collections

Captain John M. Vanderbilt first charted the reef that was later named for him. Vanderbilt worked for the Northwest Trading Co. and ferried prospectors and miners to Juneau from Sitka in the 1880s.

The first word of the *Princess Sophia's* grounding on Vanderbilt Reef 40 miles north of town came via wireless message from the ship to the Navy communications office in Juneau at 2:15 a.m. on Thursday, the 24th. The *Sophia* was due to dock in Juneau that morning and pick up another 19 passengers for the voyage south to Vancouver. Instead, the Canadian steamer, running late, followed an ill-fated course and ended up stuck high on the reef in Lynn Canal.

Vanderbilt Reef was a well-known hazard, named for the first mariner to chart it, Capt. John M. Vanderbilt, who worked for the Northwest Trading Co. hauling miners and supplies to Juneau in the 1880's. On the stormy night of the *Sophia* grounding, the reef's day marker and bell were probably impossible to see or hear until it was too late.

Postcard, by Winter and Pond. Image courtesy of the Juneau-Douglas City Museum, JDCM 88.11.001

Winds whip the sea around the Princess Sophia *as she rests precariously on Vanderbilt Reef. Rough seas thwarted multiple chances to evacuate the passengers and crew to smaller rescue vessels.*

Postcard, by William H. Case, 1918. Image courtesy of the Juneau-Douglas City Museum, JDCM 93.20.007

The SS Princess Sophia *is perched on Vanderbilt Reef after running aground at night in a snowstorm. The 353 passengers and crew were lost when the ship slipped off the reef before they could be evacuated.*

"The records of the light-keepers (on Sentinel Island and Eldred Rock) show that at 2:10 a.m., when the *Princess Sophia* struck Vanderbilt Reef, there was a strong northwest gale blowing and a blinding snowstorm, obscuring every landmark," Gov. Thomas Riggs told the public and press in a statement.

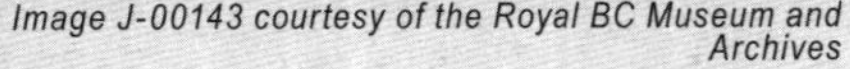

Image J-00143 courtesy of the Royal BC Museum and Archives

Capt. Leonard Locke was in command of the 245-foot Princess Sophia *and its crew of 18 officers, stewards and other staff. Locke had many critics for his decision to wait for a storm to abate and not evacuate his ship during the 40 hours it was perched on the reef. A formal investigation by the Canadian government concluded the sinking was due to natural causes.*

Photo ASL-P0I-1970 courtesy of Alaska State Library Historical Collections

Gov. John Riggs coordinated the territorial government's response to the aftermath of the sinking, personally searching beaches and coves for victims' bodies and responding to dozens of letters and telegrams inquiring about loved ones or friends. Riggs cited the Princess Sophia *disaster in lobbying the federal government for improvements to navigation in Alaska waters.*

Capt. Leonard Locke notified the ship's owners in Vancouver and authorities in Juneau that the *Sophia* and its passengers were securely lodged on the reef and the collision had caused no major damage to the doubled-hulled ship. Capts. John Leadbetter and W.C. Dibrell of the *Cedar* and Capt. James J. Miller of the *King and Winge*, were the first rescue vessels to reach the scene on the 24th. They offered to transfer passengers, but Locke decided they would be safer remaining aboard the steamer until the weather abated.

"The bow of the vessel was resting on Vanderbilt Reef near the highest point of the reef," Riggs reported. "On her starboard, or lee, side there lay a part of the reef and on her port, or windward, side was another jutting rock. The wind carried the seas around under her stern and over the reef on the lee side. It would have been impossible to launch a lifeboat either from the bow or from the stern with safety, unless the weather had moderated."

The *Sophia* was equipped with an adequate number of lifeboats and rafts, a precaution taken after the sinking of the *Titanic* in 1912. But until the weather abated in Lynn Canal that October 25th, an evacuation would have to wait, as well as any attempt to refloat the *Sophia* herself.

The record of shipwrecks in Lynn Canal shows two instances of a grounded vessel being refloated successfully -- the *Princess May* after it struck Sentinel Island in August 1910 and the following year when the *Sophia* grounded on Sentinel Island and floated off in two hours. Just days before the *Sophia* accident, the southbound steamer *Alaska* grounded in British Columbia and was pulled off without major damage.

But the weather worsened on Oct. 25th, and the *Cedar* and *King and Winge* were forced to end their watch at Vanderbilt Reef and seek refuge in nearby harbors. At 4:50 p.m., as high tide approached, the *Sophia* radioed a call for help, as it was foundering on the reef. The *Cedar* and Capt. Leadbetter attempted to reach the *Sophia* from Sentinel Island but were turned back by the storm.

"For God's sake hurry," the last wireless message from the ship read at 5:20 p.m. "Water coming in room."

On Saturday, the 26th, the *Alaska Daily Empire* front page headline declared, "*Princess Sophia* Sinks and 350 Souls Probably Perish."

The recovery effort then got underway as Juneau and the rest of Alaska and the Yukon absorbed the shocking news.

THE ALASKA DAILY EMPIRE

"ALL THE NEWS ALL THE TIME"

VOL. XII, NO. 1350. JUNEAU, ALASKA, SATURDAY, OCTOBER 26, 1918. MEMBER OF ASSOCIATED PRESS PRICE TEN CENTS

PRINCESS SOPHIA SINKS AND 350 SOULS PROBABLY PERISH

TERRIFIC STORM DRIVES PRINCESS SOPHIA OVER REEF, ON WHICH SHE HAD SPENT TWO DAYS, AND SHE SINKS WITH ALL ON BOARD; EMPTY LIFE-BOATS INDICATE THAT CHANCE THAT ANY SURVIVE THE DISASTER IS VERY REMOTE---ONE BODY FOUND

The Canadian Pacific passenger liner Princess Sophia sunk at sometime between 8 o'clock last night and 7 o'clock this morning, and in all probability every one of the 343 souls on board met watery graves in Lynn Canal. The only possibility that some of those on board were saved is the chance that life boats were launched during the night and reached shelter with their human cargoes. This possibility is regarded as remote by those familiar with the storm that was raging all last night in Lynn Canal.

The fateful message bringing the news of the greatest disaster that ever has occurred in northern waters was received at Juneau at 9:25 o'clock this morninig. It came from the United States Lighthouse Tender Cedar, and it held out no hope for those on the ill-fated Canadian liner. The message, referring to the Princess Sophia, said:

"Driven over reef during night. Only masts showing. No survivors."

With the King and Winge the Cedar immediately began the search for the passengers—the living, if any, the dead if none survived.

LITTLE HOPE EXISTS THAT ANY LIVE.

This afternoon a wireless dispatch says the Cedar had picked up four empty and capsized life boats from the Princess Sophia, and the King and Winge one. The King and Winge had recovered one body—a woman, unidentified.

The Princess Sophia ran ashore on Vanderbilt reef, four miles from Sentinel Island, at 2 o'clock Thursday morning. Since that time the weather has been too rough to transfer passengers. Boats have been lying by all the time. Yesterday the storm became terrific. Boats that were lying by sought shelter at night. At 8 o'clock Capt. Locke of the Princess Sophia sent a wireless dispatch to General Agent Lowle which said the passengers' conditions were normal, that the vessel was not taking water, but that it was too rough to transfer passengers.

That is the last that was heard from the scene until 7 o'clock this morning when the Cedar, which had been compelled to seek shelter, wired that she was leaving for the Princess Sophia. About two hours later came the fateful wireless dispatch saying that the vessel had sunk and that there were no survivors.

The circumstance that the Princess Sophia was blown over the reef leads to the conclusion that the climax of the disaster came at high tide, about 4:30 this morning.

The story of the last hours of the doomed vessel and her hundreds of human souls will probably never be told.

The disaster is probably the worst that ever has occurred in northern waters. There seems hardly a chance that a single life has been saved to tell the tale.

RECOVER LIFE BOATS.

At 3:20 this afternoon the customs house received a message from the lighthouse tender Cedar that four capsized boats had been picked up. The King and Winge picked up one unidentified body of a woman. The message said the boats were still cruising around Sentinel and Lincoln islands in the hopes of finding some survivors.

EVERY AVAILABLE BOAT TO THE RESCUE.

Every available boat at Juneau and vicinity has been sent to the scene of the disaster. The Princess Alice will be due here at 8 o'clock, and she will leave immediately. Among those who will leave on the Princess Alice for the wreck will be Gov. Thomas Riggs, Jr.

LIST OF THE PROBABLE DEAD ON PRINCESS SOPHIA

W. S. Amalong and wife
Geo. L. Shelpeth
W. Harper and wife
F. W. Elliott
Mrs. Al Winchell

S. J. Baggerly and wife
E. M. Bell and wife and two children
John Zaccarelli

(Continued on Page Two)

PRESIDENT AND ALLIES ARE IN UTMOST ACCORD

Allied Government Believed Have Approved Everything President Wilson Has Done.

ALL PAPERS DO APROVE

London and French Papers Say Final Reply Met Situation Fully and Satisfactorily.

(Continued on Page Eight)

HUNS' ALLIES WILL ALL QUIT AND SURRENDER

Austria-Hungary and Turkey Will Fight No More, but Speedily Surrender Unconditionally.

A NEW NATION IS BORN

October 21 Marked Birth Day of New Nation; the First Victory of Her Troops in Battle.

(Continued on Page Eight)

PRESIDENT ASKS PEOPLE TO VOTE FOR DEMOCRATS

Wilson Wants Democratic Congress as Endorsement of Administration in War.

ALLIES WIN IN HARD FIGHTING AT VALENCIENNES

Severest Battle of War Wages in Present Conflict; British Take 8,400 Prisoners.

AMERICANS STILL WIN

Yankees Improve Positions on Meuse; Italians Also Making Gains in South Europe.

(Continued on Page Two)

GERMANS DEMAND REPUBLIC AND KAISER TO QUIT

Mobs Storm Reichstag Asking for Abdication of Kaiser and Formation of Republic.

THE PAPERS SAY PEACE

German Publications Regard Wilson's Note as Drastic but See That Peace Must Come

(Continued on Page Eight)

Photograph by Paul J. Sincic. Image courtesy of the Juneau-Douglas City Museum, JDCM 89.21.013

The front page of the Daily Alaska Empire *on October 26, 1918. The sinking of the* Princess Sophia *set off weeks of recovery efforts to collect the remains of 353 victims and return them to families in the U.S. and Canada.*

Part III

Recovery and Rumors

Rumors attend any tragedy, and tales flew rapidly from the *Sophia's* sinking. Some people claimed they could still hear a piano playing from the sunken ship. Others told ghost stories about the makeshift morgue set up in downtown Juneau to process the remains of victims. Sacks or strongboxes of gold frequently figured in other yarns.

One Juneau family had a vivid memory of the wreck, as recalled by Emmett M. Bothelo, in a 1959 letter written by his wife. Emmett was a young boy in 1918 living with parents and a brother in Juneau.

Photo courtesy of Bothelo family

Clara Conway Bothelo poses for a photo in Dawson City around 1905-06. She was living in Juneau in 1918 with her family and expecting a visit from a friend, Cynthia Perkins, who was aboard the Princess Sophia. *Early on the morning of the grounding, Clara awoke with a premonition of the wreck that would prove all-too true. She is pictured here with her two sons, Frank (left) and Emmett.*

"On the morning of the tragedy my father, mother, brother and I lived at 3rd and Franklin in Juneau," Emmett's story goes. "About 5:30 a.m. my father called my brother and I downstairs. My mother was sitting straight up in bed and looked as if she had been in a trance. A friend from Dawson Mrs. (Cynthia) Perkins and her son, (William) Sharon, had appeared to her in a dream. My mother had awakened and screamed and related her fears to us that Mrs. Perkins and Sharon had been in a shipwreck.

"My brother and I then dressed and went down to the Customs house which was a block away so that we could check the passenger list. As we stepped out of the house, we met Mr. Whittier, Assistant U.S. Customs Inspector, and he asked me if I had heard the 'news'. I said 'Just a minute' and proceeded to explain what my mother had told us about the ship sinking. We walked to the office with him and checked the passenger list. Mrs. Perkins and her son were at the head of the list."

The list of passengers was soon eclipsed by the list of bodies recovered by a fleet of boats dispatched to the scene by Governor Riggs and Canadian Pacific headquarters in Vancouver.

Photo courtesy of Alaska State Library ASL-Juneau Ferries-03

The Lone Fisherman *normally served as a ferry between Douglas Island and Juneau. The boat and Capt. Charles Duffy helped in the recovery of bodies.*

Photo courtesy of Alaska State Library ASL-MS10-4-01-22-064

The U.S.S. Peterson *joined the search from Haines, where it was stationed with the U.S. Army under the command of Capt. Cornelius Stidham.*

The *Cedar* and the *King and Winge* were joined by the *Estebeth*, a 65-foot mail boat, and master James P. Davis; the *Amy*, a 65-footer skippered by Edward McDougal; the *Lone Fisherman* and Charles Duffy; the *Peterson*, an Army patrol boat based in Haines and captained by Cornelius Stidham; and the *E.A. Hegg*, a 46-foot boat piloted by Niels P. Madsen.

Photo courtesy of Alaska State Library

The U.S.S. Lighthouse Tender Cedar *was among the first boats to reach Vanderbilt Reef from nearby Sentinel Island. Capts. John Leadbetter and C.W. Dibrell were in command of the 100-foot tender.*

Photo courtesy of Alaska State Library ASL-MS10-4-03-105

The Estebeth, *a 65-foot mail boat that served in Southeast Alaska, had Master James P. Davis at the helm.*

Photo courtesy of Alaska State Library

At 100 feet in length, the King & Winge *could have taken on many of the 350 passengers aboard the* Princess Sophia. *But Capt. James Miller's offer to evacuate them was not accepted by Capt. Leonard Locke of the* Sophia.

Photo courtesy of Alaska Electric Light & Power

Edward McDougal skippered the 65-foot Amy *in the search for victims of the shipwreck.*

Riggs issued a proclamation ordering flags to half-staff across the territory and asking churches to hold memorial services. He then joined the recovery effort aboard the *King and Winge* on Oct. 27.

"Upon arriving at Young's Bay at eight a.m. we saw gas boats picking up bodies...By ten o'clock we had put thirty bodies on board (the *King and Winge*)," Riggs reported. "All the bodies found had on life preservers and were floating in a thick scum of oil...I saw no evidence of any persons having been lashed to life rafts, although five were picked up near the head of Shelter Island near a wrecked raft."

This fleet scoured beaches along Lynn Canal and by early November brought 180 bodies to a mortuary at the C.W. Young waterfront store in downtown Juneau. All the victims were coated in diesel oil from the *Sophia.* They had to be stripped and scrubbed with gasoline with the help of Juneau volunteers. Other friends and relatives and Northwest Mounted Police traveled from Whitehorse and Skagway to help in identifying the victims.

Photo courtesy of Brian Wallace

Henry Cropley served aboard a vessel that joined the rescue attempt on the Princess Sophia. *He is pictured here at 28 in the uniform of the Revenue Service-Coast Guard.*

Diesel oil spilled in the accident spread widely, killing "thousands and thousands" of ducks and other seabirds, reported Game Warden John Land, Jr.

An article in the *Daily Alaska Empire* on October 29, headlined "Volunteers Are Accomplishing Much Good Work," reported that Juneau residents volunteered to handle the bodies of *Princess Sophia* victims being brought to town and prepared for burial, "which is little short of marvelous, considering the shortage of experts and man-power, depleted by the war."

Photo courtesy of Joel Bennett

Vanderbilt Reef exhibits a scar, possibly from a ship striking it.

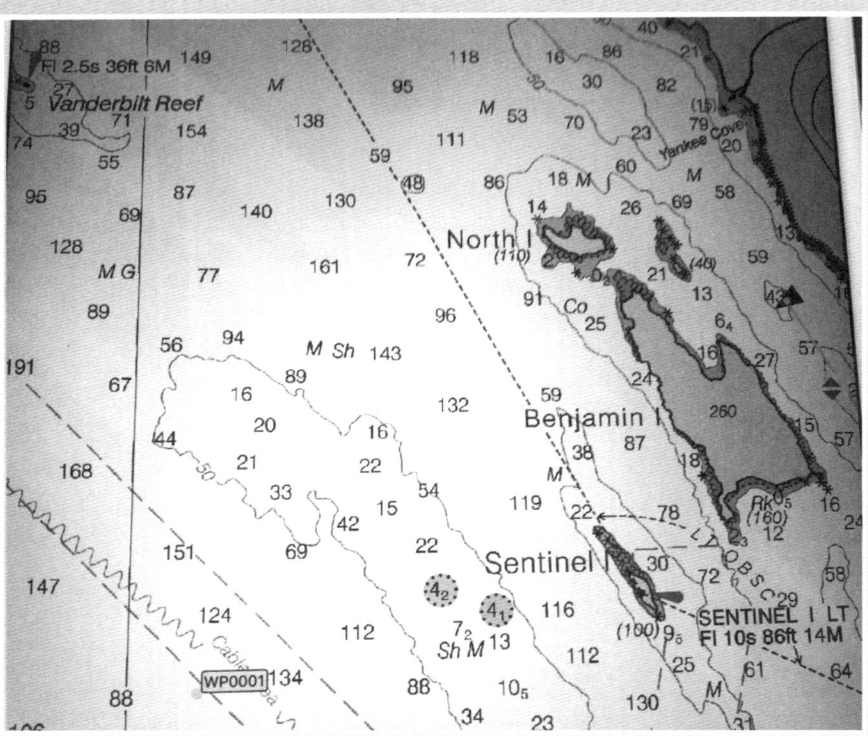

This nautical chart shows Vanderbilt Reef (at left) in Lynn Canal. Sentinel and Benjamin Islands (at right) offered shelter for rescue boats that responded to the plight of those aboard the Princess Sophia. *Sentinel Island hosts a lighthouse commissioned in 1902.*

As the remains arrived by boat, the article says, they were processed through the undertaker parlors in town under the supervision of Edward Provence. They were then transferred to the warehouse, which was guarded day and night, under the leadership of Dan Sutherland and Harry Pryde.

The article reported that volunteers next searched for personal effects under the supervision of Harry Fisher, Ed. Hurlbutt, William Geddes, and Mort Truesdell. Clothing and shoes were put into a container for use by the Red Cross.

The next step was cleaning the oil-stained bodies with gasoline. R. Morres supervised this chore, assisted by Oak Olsen, Walter Bathe and J. L. Gage. Men handled the male victim's bodies, and women handled the women and children, the *Empire* reported.

George W. Mock, Charles W. Cartey, Charles W. Hawkesworth, A. W. Rhodes, and other volunteers performed embalming of the bodies.

At another location, valuables and personal effects were marked and identified by F.W. Lowie, general agent for Canadian Pacific; Maurice S. Whittier, U.S. Customs Office; Guy McNaughton, Harry Lucas and J.R. Willis, of B.M. Behrends Bank; J.W. Ball, U.S. District Court clerk's office; and Charles Harland of the Territorial Treasurer's Office.

The article singled out John C. McBride of the C.W. Young Co. for his contributions, especially for recruiting embalmers and additional coffins from Ketchikan, Skagway and Seattle.

"I feel the disaster probably as much as any man in Alaska as there were on board fully fifty people with whom I was acquainted, many of them intimately, and whom I regarded with deep affection," Riggs wrote. "The loss sustained by the Government and by the Territory in the death of Mr. (John) Pugh, the Collector of Customs, is one that it is hard to become reconciled to."

This building on Front Street in downtown Juneau served as a morgue for about 180 victims of the sinking. Over about two weeks, the bodies recovered from beaches and bays, and often buried in snowdrifts, were identified, cleaned and prepared for shipping south and burial. Twenty-one victims were buried in Evergreen Cemetery. Funeral costs were reimbursed by Canadian Pacific Steamship Co., but the company fought and won an ensuing legal battle over liability for the sinking.

Pugh had traveled to Skagway to help passengers board the *Sophia*, among them Walter Harper and Frances Wells Harper, his wife of seven weeks. Known widely as the first man to reach the summit of Denali in the Hudson Stuck expedition of 1913, Harper, an Athabascan, was leaving Alaska for medical training and possibly to return to the hospital in Tanana, where he worked with Archdeacon Stuck and met Wells, who was a nurse.

By chance, among the first bodies recovered by searchers and brought to the C.W. Young mortuary were those of Pugh and the Harpers.

On Nov. 4, Archdeacon Stuck wired Riggs from Ft. Yukon asking about Walter Harper's possessions: "Have reason to believe Walter Harper had something like eight hundred dollars in currency in a money belt about his waist belonging to himself and wife. If found upon him should be held for his mother."

Image J-00137 courtesy of Royal BC Museum and Archives

A dockside facility in Vancouver, B.C., received the remains of passengers from the Princess Sophia *sinking. These victims were lacking positive identification after volunteers in Juneau had recovered, cleaned and prepared them for burial. Ironically, Vancouver was in the midst of a citywide celebration when the bodies arrived on Nov. 11, 1918, the armistice of World War I.*

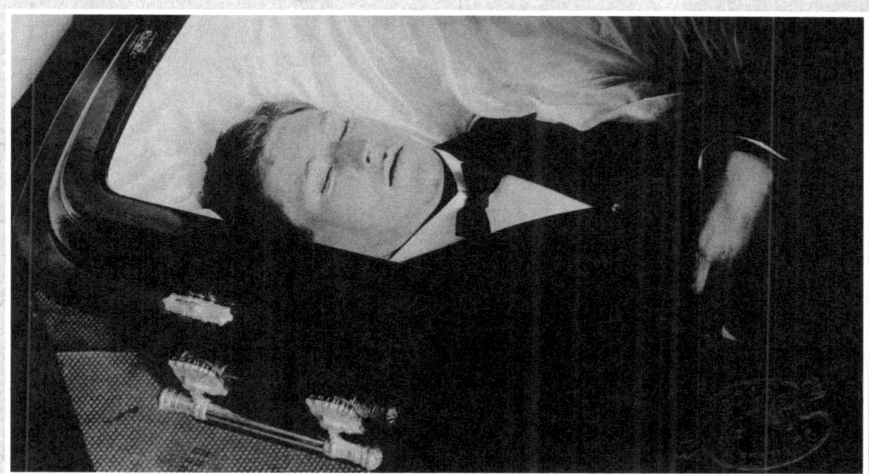

Image J-00135 courtesy of Royal BC Museum and Archives

A victim of the Princess Sophia *sinking awaits identification in Vancouver.*

Riggs replied on Nov. 6: "Walter Harper had two hundred cash, one fifty-dollar Liberty Bond, one Wells Fargo draft for one hundred, gold watch and ring. Mrs. Harper fifty-five dollar cash and jewelry. Walter may have had money on deposit with purser, whose safe not found. All estates will have to go through probate court before turning over to heirs."

As news of the sinking spread, Riggs received dozens of inquiries from relatives of the victims. H.E. Seneff of Fairbanks wrote on Nov. 25 to Riggs about his son, Edgar:

"Edgar was our only child and of course was very dear to us. He was eighteen years old last May. At the time of his fate he was on his way outside to attend a Naval School, as he wished to get in that branch of the service. Up to this writing his body has not been found."

Riggs had ordered the formal search for bodies suspended on Nov. 20, but the Canadian Pacific Steamship Co. offered $50 for each body found after that. Divers had been able to recover one body from the submerged wreck, but others were probably still on board, Riggs' secretary informed Seneff.

Eventually, divers did recover more bodies from the wreck, in addition to sacks of mail and 230 pounds of gold.

On Nov. 14, the bodies of 21 victims were interred at Evergreen Cemetery, including those of the Harpers. Walter, 25, and Frances, 29, were buried side-by-side.

Photo courtesy of John Greely

The burial site of Walter and Frances (Wells) Harper is located in Juneau's Evergreen Cemetery. They were married just seven weeks before boarding the Princess Sophia *in Skagway on a journey from Fort Yukon to Frances' home in Pennsylvania. The Harper family and the Episcopal Church of Alaska have helped to maintain the gravesite, including the addition of a metal nameplate.*

PART IV
CONCLUSION

The bodies of 40 passengers were never found, including that of Capt. Locke. His actions were second-guessed by many, especially by families of the victims.

Mrs. Clarence S. Verrill of Vancouver, B.C., wrote to Riggs on Nov. 4 about her late husband's belongings. He was to be buried in Juneau with other victims.

"I am sorry," Riggs wrote in reply, "that it is reported around Vancouver that the passengers were not taken off the *Princess Sophia* owing to the desire not to incur a bill for salvage. I do not consider that this is the truth, as Captain Locke had made all arrangements to take off the passengers as soon as the weather should have moderated, as the indications were for fair weather...That there was an error in judgment on his part is manifest, but he himself lost his life on account of such error."

M.J. O'Connor, a businessman and former mayor of Douglas, lashed into Riggs: "We commend the work of the governor in assisting in the recovery of bodies of the victims of the disaster, but believe that he should have suspended judgment until an official government investigation was made by men competent to pass on questions of navigation and the duties of master mariners...I may be a landlubber, but I am certain that no vessel can be safe on top of a rock and especially in the waters of storm-swept Lynn Canal in the month of October."

An official inquiry commenced in Vancouver/Victoria and moved to Juneau in early 1919. No liability for the wreck was found in those hearings or subsequent litigation.

For Riggs, the focus moved to improving navigational safety in Alaska waters. He had an ally in the U.S. Secretary of Commerce, William C. Redfield, whose agency oversaw the newly minted U.S. Coast Guard.

In a letter just three days after the sinking, Redfield told Riggs that Congress had shown "hideous neglect" in providing proper safeguards in Alaska waters. In reply, Riggs warned Redfield that funds were needed not only for aids to navigation but also to carry on additional surveys of marine hazards, and that "until adequate funds are forthcoming, we may expect repetitions of this last terrible calamity."

More and better navigation aides would follow, spurred on by the onset of World War II and greatly increased ship traffic in the North Pacific and Southeast Alaska. Vanderbilt Reef got a lighted buoy soon after the *Sophia* went down.

History has all but overlooked the tragic sinking of 1918 and Juneau's role in coping with the disaster on its doorstep. It was up to a town of less than three thousand to take on the sad task of finding the scattered human remains on the islands and beaches of Lynn Canal and transport them to a temporary morgue, where they were cleaned of oil so thick they were hard to recognize. Everyone who could help with these tasks did so, knowing that it was up to them to bring a last moment of human dignity to the victims of the sinking.

The people of Juneau were heroic and courageous in doing what had to be done. This book is dedicated to those courageous heroes.

Photo courtesy of John Greely

A plaque commemorating the centennial of the 1918 sinking, the worst in North Pacific coast history, has been erected at Eagle Beach State Park north of Juneau. The project by the Pioneers of Alaska Igloo 6 overlooks the waters where the Princess Sophia *rests today.*

* SINKING OF THE S.S. PRINCESS SOPHIA * *100TH ANNIVERSARY *

The members of the Thirtieth Alaska State Legislature honor the memory of those lost in the sinking of the S.S. Princess Sophia a century ago, on October 25, 1918. We honor the steadfast efforts of the Alaskans who did all they could to rescue the victims and respond to the tragedy.

In a blinding winter snow storm the night of October 23, 1918, the S.S. Princess Sophia left Skagway three hours behind schedule. Steaming at near maximum speed, with over 350 souls aboard, she ran aground on Vanderbilt Reef at 2:00 in the morning. Mariners aboard the ship tried valiantly to keep death at bay. The Sophia's captain and crew, and those who responded to the call for aid, braved the gales to save lives in peril.

Local mariners turned out in force, attempting to rescue passengers. Even as the snow abated, the weather remained too rough to lower lifeboats or bring small craft alongside to evacuate passengers. The U.S. Lighthouse Service tender Cedar arrived the evening of October 24, but was unable to help in the wind and tides. Rescue was close, but never close enough.

The Sophia remained stuck fast for over 40 hours, pounded by the wind and waves, until the evening of October 25th, when the worsening storm pushed her off the reef. When the Sophia sank, every single person aboard perished in the fuel-blackened sea. Valiant would-be rescuers who returned to port without rescuing a single survivor carried the pain for many years. Alaskan boaters went back to the water and recovered the dead.

A century ago, Juneau was a small town of 3,000. When the bodies arrived, the territorial capital received a deluge of horrors equal to more than 10% of its population. The community was suffering from an epidemic of spanish flu. In the shadow of the 1918 armistice that ended one of the world's greatest tragedies, Alaskans did what had to be done. With solvent and rags, they cleaned oil from the corpses and returned to the dead their human dignity. Afterward, they completed the grim work by returning the dead to their bereaved families – in coffins brought in from all over Southeast because no single community had the numbers needed. Corpses continued to be found for three months after the sinking.

In hopes the sinking of the Sophia will remain the greatest maritime tragedy in the Pacific North West; the members of the Thirtieth Alaska State Legislature honor the memory of the S.S. Princess Sophia and those lost at sea. We honor the efforts of the Alaskans who came to their aid.

BRYCE EDGMON
SPEAKER OF THE HOUSE

PETE KELLY
PRESIDENT OF THE SENATE

SEN. DENNIS EGAN
PRIME SPONSOR

Source Material

Part 1

Treadwell Gold, an Alaska Saga of Riches and Ruin, University of Alaska Press, Sheila Kelly

1918 Spanish Influenza in Alaska, Wikipedia

Part 2

Robert N. DeArmond, Juno's Days of Yore, Info Juneau, May 25, 1991, newspaper clipping

Gov. Thomas Riggs, official files, Alaska State Museum, statement "For the Press," November 1918

Aids to Navigation in Alaska History, History and Archaeology, Series No. 7, Miscellaneous Publications, Alaska Division of Parks, November 1974

Parts 3 & 4

Spirits of Southeast Alaska, James P. Devereaux

Walter Harper, Alaska Native Son, University of Nebraska Press, Mary F. Ehrlander

Various correspondence, Gov. Riggs official files, Alaska State Museum

---

Photo Credits

End of the Treadwell, CBJ Museum files, April 22, 1917

Princess Sophia Stranded on Vanderbilt Reef, Oct. 24, 1918, Winter & Pond JDCM88_11_001

Princess Sophia on Vanderbilt Reef, Winter & Pond, JDCM 88_11_002

"*Princess Sophia Sinks and 350 Souls Probably Perish*", Alaska Daily Empire front page, Oct. 26, 1918, JDCM, 89_21_013

Princess Sophia at dockside Juneau waterfront, 1918, JDCM 98_20_088

Princess Sophia at dockside Juneau waterfront, 1918, JDCM 2015_29_007

Rescue and recovery boats *Cedar, Estebeth, Peterson, King and Winge*, courtesy Puget Sound Maritime Historical Society

Capt. John M. Vanderbilt, courtesy Alaska State Museum

Current views of *Vanderbilt Reef*, courtesy Joel Bennett